Extreme Dinosaur Hunt

by Ruth Owen

Consultant:
Dougal Dixon, Paleontologist
Member of the Society of Vertebrate Paleontology
United Kingdom

New York, New York

Credits

Cover, © Edson Vandeira/Alamy and © Anton Rodionov/Shutterstock; 2–3, © Kazuhiro Nogi/Getty Images; 3B, © Krasowit/Shutterstock; 4, © Aurora Photos/Alamy; 5T, © Cosmographics; 5B, © U.S. Antarctic Program, National Science Foundation; 6, © Murray Foubister/Creative Commons; 7, © Mike Lucibella/U.S. Antarctic Program, National Science Foundation; 8TR, © Andy Sajor; 8C, © Radius Images/Alamy; 8B, © Artic_Photo/Shutterstock; 9, © Yuki Takahashi/U.S. Antarctic Program, National Science Foundation; 10–11, © Mike Lucibella/U.S. Antarctic Program, National Science Foundation; 12, © Philip Currie; 13, © Mike Lucibella/U.S. Antarctic Program, National Science Foundation; 14–15, © Eva Koppelhus; 16, © James Kuether; 17, © Jonathan Chen/Creative Commons; 18, © U.S. Antarctic Program, National Science Foundation; 19, © James Kuether; 20T, © Peter Rejcek/U.S. Antarctic Program, National Science Foundation; 20B, © Jack Green/U.S. Antarctic Program, National Science Foundation; 21, © Kazuhiro Nogi/Getty Images; 22T, © Warpaint/Shutterstock; 22B, © Mike Lucibella/U.S. Antarctic Program, National Science Foundation; 23T, © Marcio Jose Bastos Silva/Shutterstock; 23C, © Thomas Barrat/Shutterstock; 23B, © benedek/Istock Photo.

Publisher: Kenn Goin
Senior Editor: Joyce Tavolacci
Creative Director: Spencer Brinker
Image Researcher: Ruth Owen Books

Library of Congress Cataloging-in-Publication Data

Names: Owen, Ruth, 1967– author.
Title: Extreme dinosaur hunt / by Ruth Owen.
Description: New York, New York : Bearport Publishing, [2019] | Series: The dino-sphere | Includes bibliographical references and index.
Identifiers: LCCN 2018053279 (print) | LCCN 2018055217 (ebook) | ISBN 9781642802597 (Ebook) | ISBN 9781642801903 (library)
Subjects: LCSH: Dinosaur tracks–-Antarctica—Juvenile literature. | Dinosaurs—Antarctica—Juvenile literature.
Classification: LCC QE861.6.T72 (ebook) | LCC QE861.6.T72 O94 2019 (print) | DDC 567.9—dc23
LC record available at https://lccn.loc.gov/2018053279

Copyright © 2019 Ruby Tuesday Books. Published in the United States by Bearport Publishing Company, Inc. All rights reserved. No part of this publication may be reproduced in whole or in part, stored in any retrieval system, or transmitted in any form or by any means, electronic, mechanical, photocopying, recording, or otherwise, without written permission from the publisher.

For more information, write to Bearport Publishing Company, Inc., 45 West 21st Street, Suite 3B, New York, New York 10010. Printed in the United States of America.

10 9 8 7 6 5 4 3 2 1

Contents

Welcome to Antarctica! 4
The Coldest Place on Earth 6
Building a Camp 8
Dinosaur Mountain 10
Extreme Fossil Hunting 12
Lift Off! 14
New Dinosaurs 16
A Different Time 18
The Hunt Goes On 20

Glossary 22
Index 24
Read More 24
Learn More Online 24
About the Author 24

Welcome to Antarctica!

A plane lands in icy Antarctica (ant-AHRK-ti-kuh).

A group of **fossil** hunters climbs from the plane.

The **scientists** are ready to hunt for dinosaur bones!

Antarctica is a huge area of icy land. The only people who live there are scientists.

Antarctica

The Coldest Place on Earth

In Antarctica, it's colder than the inside of a freezer!

Powerful winds blow snow across the icy land.

To keep warm, fossil hunters wear layers of clothing.

They also wear snow pants and thick, hooded jackets.

hooded jacket

snow pants

In Antarctica, people wear sunglasses. The brightness of the snow and ice can make it hard to see.

Building a Camp

Before they look for dinosaurs, the scientists set up camp.

They put up small tents.

The fossil hunters use a small stove for cooking and melting ice for water.

stove

They build walls with ice blocks to protect the tents from the wind.

building a wall

cutting ice blocks

Dinosaur Mountain

Fossil hunters search for dinosaur remains in rock.

However, the rocky land is buried under thick ice.

To find fossils, the scientists look on a mountain where there's less ice.

They fly up the rocky slopes on a helicopter.

The ice sheet that covers Antarctica is more than 1 mile (1.6 km) thick!

Extreme Fossil Hunting

For hours, the scientists carefully look for dinosaur bones.

When they spot a fossil, they dig it up.

jackhammers

In icy Antarctica, the rock is very strong and hard.

The fossil hunters use special machines to get to the fossils.

The fossil hunters use tools called jackhammers and rock saws.

rock saw

Lift Off!

Sometimes, the scientists dig up a fossil inside a hunk of rock.

To move it, they drag the rock onto a strong net.

They clip the net to the helicopter.

Then, the helicopter carries it back to camp.

Fossil hunters often work late at night. At certain times of the year in Antarctica, it never gets dark!

New Dinosaurs

In 1986, scientists discovered the first dinosaur in Antarctica.

They named it *Antarctopelta*.

Antarctopelta was a plant-eater. It had **armor** and spikes.

spikes

armor

Antarctopelta
(ann-TARK-toh-PEL-tah)

Scientists found bones from a meat-eating dinosaur in 1991!

They named it *Cryolophosaurus*.

It was 20 feet (6 m) long and weighed as much as a big horse.

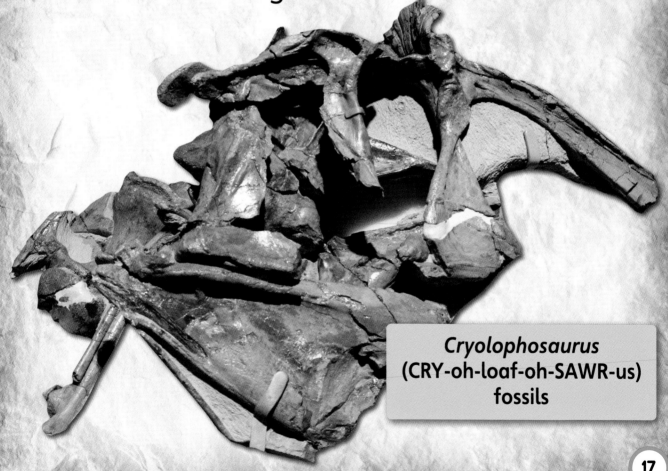

Cryolophosaurus
(CRY-oh-loaf-oh-SAWR-us)
fossils

A Different Time

How did dinosaurs survive in icy Antarctica?

They lived millions of years ago when Antarctica was warmer than it is today.

It had forests and was home to many kinds of animals.

Fossil hunters found fossilized plants in Antarctica. The plants were alive at the same time as *Cryolophosaurus*!

The Hunt Goes On

When an **expedition** is over, the scientists fly home with their fossils.

boxes of fossils

The fossils are taken to **museums** and studied.

Hunting for dinosaurs in icy Antarctica is tough work.

However, it's the perfect job for extreme fossil hunters!

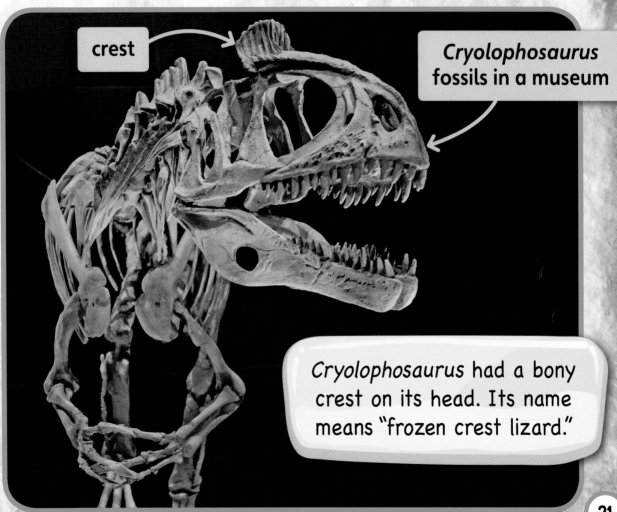

crest

Cryolophosaurus fossils in a museum

Cryolophosaurus had a bony crest on its head. Its name means "frozen crest lizard."

Glossary

armor (AR-mur) a tough covering that protects an animal or a person's body

expedition (ek-spuh-DISH-uhn) a long trip taken for a special reason, such as fossil hunting

fossil (FOSS-uhl) the rocky remains of an animal or a plant that lived millions of years ago

museums (myoo-ZEE-uhmz) buildings where interesting objects, such as fossils and art, are studied and displayed

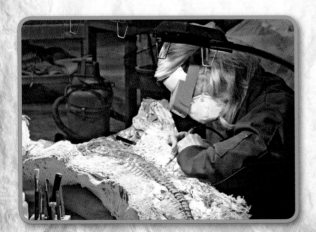

scientists (SYE-uhn-tists) people who study nature and the world

Index

Antarctica 4–5, 6–7, 11, 13, 14, 16, 18, 21
Antarctopelta 16
armor 16
camps 8–9, 14
clothing 7
Cryolophosaurus 17, 18–19, 21
dinosaurs 5, 8, 10, 12, 16–17, 18–19, 21
fossils 4–5, 10, 12–13, 14, 16–17, 18, 20–21
helicopters 10–11, 14–15
museums 20–21
tools 12–13

Read More

Ganeri, Anita. *Introducing Antarctica (Heinemann First Library).* North Mankato, MN: Heinemann-Raintree (2014).

Owen, Ruth. *Fossils: What Dinosaurs Left Behind (The Dino-sphere).* New York: Bearport (2019).

Learn More Online

To learn more about fossil hunters and dinosaurs, visit www.bearportpublishing.com/dinosphere

About the Author

Ruth Owen has been developing and writing children's books for more than ten years. She first discovered dinosaurs when she was four years old—and loves them as much today as she did then!

3 1814 00330 8175